Two Lessons on
Animal and Man

Gilbert Simondon

Two Lessons on Animal and Man

Introduction by Jean-Yves Chateau

Deux leçons sur l'animal et l'homme
- published by Ellipses -
Copyright 2004, Édition Marketing S.A.

Translated by Drew S. Burk
as *Two Lessons on Animal and Man*

First Edition
Minneapolis © 2011, Univocal Publishing

Published by Univocal
123 North 3rd Street, #202
Minneapolis, MN 55401

Thanks to Nathalie Simondon,
Dominique Simondon and
the Association pour le musée Jean de La Fontaine.

Designed & Printed by Jason Wagner

Distributed by the University of Minnesota Press

ISBN 9781937561017
Library of Congress Control Number 2011945799

Table of Contents

Translator's Note

It is my great pleasure to have the opportunity to be involved with one of the first book-length English translations of an extraordinary thinker and philosopher. For many, Gilbert Simondon is an unheard of landscape of philosophical inquiry. For other thinkers such as Gilles Deleuze, his work on individuation is essential for the task of moving outside anthropocentric conceptions of identity formation and humanity's relationship to the technical universe. In this collection of early lectures, the reader gets a glimpse into Simondon's understanding of the history of philosophical discourse in regards to the human, the animal, and the vegetal.

Drew S. Burk

Introduction

The following text by Gilbert Simondon is comprised of two lessons serving as an introduction to an annual course of general psychology (which he taught until 1967) addressed to first year humanities students at the University with their sights set on an undergraduate degree in philosophy, psychology, or sociology.

The Challenge for Psychology

Psychology is a discipline, an order of research and teaching, whose determination of the object poses the problem of knowing what the relations between man and animal are: is psychology merely interested in man or does it have an interest in animals as well? The answer provided by the existence of an "animal psychology" within the technical division of labor of teaching and research, certainly does not resolve by itself this problem but ballasts it from an institutional weight: even if there are differences between human and animal psychology (which not every psychologist would perhaps agree with), the utilization of the same term "psychology" seems to imply that there is at least something in common between man and animal, human life and animal life. But, if one uses the same methods in psychology for studying man and animal, does this mean that they have, from a psychological point of view, something similar or essentially in common with each other? Otherwise, this could signify that what psychology is capable of grasping is neither essential to man, nor essential to animal.

Psychology traditionally studies what we could call the mind, the soul, consciousness, etc. But is there a reason for studying this in animals? In any case, this is not at all what animal psychologists do. Should they not rather study instinct? But psychology, *de facto*, studies both of them, in man and animal. It studies intelligent or instinctive behavior equally in humans and animals. It studies, from its point of view, human life and animal life.[1] The traditional distinction between intelligence and instinct, which has been elaborated first in order to oppose what characterized life and human behaviors to those of animals, does not allow differentiating the object of human psychology and that of animal psychology. Hence, that a superficial reflection such as psychology

---

1 In doing so, it has renewed itself with a tradition that goes all the way back to Aristotle and his treatise, *On the soul* (*De Anima*): the soul is "that which animates", the principle of life, whether we are speaking of the human, the animal, or the vegetable. What moves by itself is alive, what has its own principle of movement or change (or their absence) in itself and not by accident (in opposition to that which comes via technique). See *De Anima* and *Physics II*. "Aristotle included psychology within biology", Simondon says.

could presume to be founded upon a distinction between properly human behaviors and those of animals, shows rather the difficulty of distinguishing between the two. General psychology poses the problem of life, of the unity of human and animal life, and its relationship with intelligence, habit, and instinct.

It is via the study of this problem that Simondon envisions, in the first paragraph of the following text, introducing his annual course on general psychology. To this end, before studying the manner in which the problem is posed in current theories, he proposes studying the history (throughout a time period which goes from Antiquity to the 17th century) of the notion of animal life, which is also that of human life: one is inseparable from the other, whether it is because we cannot pit them against each other or, on the contrary, because the one is merely the opposite of the other. This historical investigation, which bears on the concept formation of contemporary psychology, is interested in showing how the determination of these concepts (and by this the determination of the fundamental object of this discipline and its methods) finds its origin in conceptions and debates in very ancient ideas, which Simondon traces back to the Presocratic thinkers. We are not dealing with a complete history about notions of human and animal life, nor are we dealing with studying them for themselves in order to represent in all their diversity and nuances the diverse related doctrines, but to make appear in a contrasted manner the principle conceptions and points on which they oppose each other as figures which represent the problem and its diverse forms.

To know whether one must distinguish between human life and animal life, to what degree, and how, is not, it seems, a question to which any science has a direct reply even though a certain number can appear to depend, in their possibility and their definition, upon an answer to such a question (as we saw with psychology). It is, however, a question almost every person has an opinion about and to which they are quite strongly attached.[2] It is a question often asked in daily life before being asked, if at all, in philosophy; and it is not only the notions of man and animal which can be a problem, it is also the terms and representations in which we pose this problem and try to resolve it ("intelligence, reason, soul, thought, conscious, body, instinct", etc.). Men suffer with difficulty if we don't share the same opinion as them, whatever it may be. This is because

---

2 Especially since everyone's conceptions of them has, in general, the earliest days of childhood as its roots: a moment where the animal and its representation had an importance, as big as it was complex, what psychology, psychoanalysis as much as common sense, know from experience.

it is the representation that we have of ourselves, of the manner in which it is appropriate to behave with others and what we can expect from them, the most fundamental values, and sometimes, what one can hope from life, even beyond it, which finds itself at stake in any conception of relations between man and animal.

The Ethical and Religious Challenge of the Problem

But, what first allows for the historical canvas that Simondon brushes to distinctly appear is the religious and moral dimension of the problem. It would have been Socrates, who, more or less, invented man, and in underlining his radical distance from everything within nature, founded a humanism based on "anthropological difference.[3] But it is the

---

3 This representation corresponds to the intellectual autobiography that Socrates exposes in Plato's *Phaedo* where he explains how disappointed he was in his youth by the naturalist research like that of Anaxagoras. Rather than looking for the natural causal chain by which things become what they are, Socrates thought that the one truly important thing was to know why one must do what one must do: if Socrates is here in his prison, it is not fundamentally due to the bones and muscles of his body (physical and physiological determinations *without which* he would not be there), but due to his belief that the Idea of Justice did not want him to do harm to the City (to which he owes everything), fleeing even an unjustified punishment. What Socrates shows is that the one thing which merits worrying about is Man: this being which has thought (*phronesis*) as a capacity to think Ideas, the highest *why* of them all.

11

eminent dignity of man that Socrates establishes in thus separating it from all other natural realities. This sentiment of an essential difference between man and animal, linked to a singular sense of man's value, is shared, starting from different principles, with the Sophists ("man is the measure of all things") as well as Plato, the Stoics, the Christianity of the Fathers of the Church, the first Apologists, and above all, Descartes. Simondon characterizes these doctrines as "ethics". Nevertheless, moral and religious values can equally lead, on the contrary, to the thesis of proximity or at least the continuity between the human and animal psyche, as in the Renaissance, Saint Francis of Assisi, and Giordano Bruno. Simondon himself underlines, as a decisive determination concerning the debate and its destiny, the vigorous moral judgment by Descartes' enemies who found his position to be "excessive, bizarre, and scandalous". But, even a representation like the one Aristotle proposes, which has as its aim to be based upon an objective observation, and which is considered by Simondon as a "generous, intelligent, non-systematic and non-dichotomic vision" (in its results if not in its principles), in the end leads to a prioritization of man in relation to other living things which, even if it "is not a prioritization for purposes of normative opposition", is obviously not axiologically neutral.[4]

---

4 All one has to do in order to be convinced is to think of the role that Aristotle

The History of Ideas and its Dialectic of the Whole

In a general way, it is visible that, during the studied time period, despite the indication of a movement of a "dialectical" ensemble of ideas, the opposing views were able to exist and come back to the forefront as well after having ceased to be the dominant view. This is the historical magnitude of the test proposed by Simondon which, even if it was not able to enrich every doctrine, is able to show the contribution of each one to a position and treatment of a problem: there is not *one* conception from Antiquity or Christianity about the question. The Presocratics and Aristotle, in Antiquity, conceived of a great continuity between man and animal; But Socrates, Plato, and the Stoics, on the other hand, underlined the singular status of man separated from the rest of nature. Within Christianity, from the beginning periods as well as the Middle Ages, there is an attachment to the devaluation of animals, completely separating them from man (at least from

---

assigns to reason, which is "the specific characteristic of man", in his morality, in the form of "practical reason" (*noûs pratikos*), of this "practical intelligence" whose virtue is *phronesis*, "prudence" (see the Nicomachean Ethics, IV). Thus the ethical scope of this specific difference is obvious, even when it would not be within a moral intention that this difference is established and, in any case, not in the intention of establishing a radical separation between man and animal even from an ethical point of view, as the Aristotelian affirmation appears to witness that there could be perhaps in certain animals a type of *phronesis*, an imitation of *phronesis*.

those who were true Christians). At the same time the devaluation of animals was occurring, the valorization of animals and their similarity to man, considering them as equals, was cultivated in a passionate manner and in both cases established themselves from a mythic conception of the animal. There is not *one* Christian conception of the relation between humanity and animality, or perhaps it is better to say that it is a *problem* for Christianity, which takes a form and particular meaning within Christianity; and in reality, there are several Christian ways of stating the problem (there are arguments for and against it which have above all a meaning to certain Christians). Nor can we say that there is *one* conception proper to modern times (in the 16th, 17th, and even the 18th century, though Simondon's investigation does not bear on it, at least in the research he used for his course) as illustrated by the conflict between Descartes' and the Cartesian's conceptions and those of the writers who countered them such as Bossuet and above all La Fontaine. Here, we can see there is a problem, which is not eternal, but which changes and reconstructs itself from one time period to another, beyond the arguments and doctrines, in grand philosophical terms.

The diverse conceptions evoked by Simondon counter each other and organize themselves as certain positions taken regarding one of the following grand questions.

- The first question is knowing whether there is continuity between man and animal or if there is an essential difference between them. The first position is that of the Presocratic "naturalists" (Pythagoras or Anaxagoras), the second one is from Plato and Socrates (A bit less frank, according to Simondon), and it is perhaps a problem to try to clearly place Aristotle somewhere regarding this point.

- If the difference between man and animal is to be recognized, then the problem becomes knowing whether we are led to what Simondon calls the split "dichotomy" isolating man from nature. This is the position of Socrates, the Stoics,[5] the first Christian Apologists and Descartes. More moderate, even if they still think there is a specific difference between man and animal, are the likes of Aristotle, Saint Augustine, but also Montaigne, Bossuet, and La Fontaine.

---

5 "They want to show that the human is a being apart from the rest of nature" (p. 53).

15

- If there is a difference between man and animal, which one is superior? What is interesting in evoking the figures of Bruno and Montaigne is showing the possibility of supporting the idea that, to a certain degree, there is a superiority of the animal.

- If it is man who is superior, the question is to know if it is due to progress in relation to the animal (which is the general position of the Presocratics such as Anaxagoras) or if there has been a degradation of man to the animal (Plato's position in the Timeaus.)[6]

---

6 Simondon is not trying to present here the totality of Plato's conception but the most significant aspects susceptible to composing the constellation of the problems and positions corresponding to our question. Simondon, whose interest in technology we know quite well, is not evoking Prometheus' myth of Protagoras (quite important for thinking about technics), which presents the creation and equipping of living things beginning with animals, and man coming in the final position, leading him to be deprived of any natural tools like all the other animals and technics, which are then granted to him, is thus presented in a problematic fashion both as something separate from all the other natural (instinctual) know-how and tools, and as a kind of (stand-in) natural know-how and tools which are properly human. But this representation (whose importance imposed itself at the heart of Western culture for centuries, and where Simondon evokes Senecca's version) is not so different than the opposing point, contrary to the Pre-Socratics, of seeing in humanity a progression in relation to animality, Simondon chooses rather to evoke Plato, the myth of the Timaeus, which, presenting the idea of animality as degradation of humanity, constitutes a figure of thought in regards to our most original problem (which

- In any event, in the end, there is no way to establish a dichotomic or hierarchical difference between men and animals but merely to affirm their homogeneity, but there still remains the problem of knowing if animals should be thought of using humans as models, which was the position of the Ancients (endowed with reason, intelligence, a rational soul, etc.), or through the counter model of Cartesianism where man is considered according to animal models. It is this latter position that will impose itself at least within the history of the constitution of contemporary psychology.

The presentation Simondon makes of Descartes[7]

---

he even qualifies as both genius and monstrous at the same time).

7 The figure of Descartes presented here is perhaps closer to a reception by certain "Cartesians" who were a bit stiff (like Malebranche), or a hostile reception, like that by La Fontaine, which he evokes with an obvious sympathy, than what Descartes himself claims, if we take into account everything he wrote. It is true that it is Descartes' *entire philosophy* which finds itself engaged here, if we want to judge it, and it is not a small effort to provide a fair idea of what Descartes said concerning this subject with as much precision and nuance as firmness. To help us in this examination, we could especially consult: *Méditations métaphysiques* VI, *Réponse aux 4 Objections* (Pléiade, p. 446 and above all p. 448-449) and *Réponse au 6ᵉ Objections*, 3ᵉ (p. 529-531); *Traité de l'homme* (above all p. 807 and p. 872-873); *Discours de la méthode* V; Lettre à Reneri pour Pollot (April 1638), A. T. II, p. 39-41; Lettre à Newcastle from 23-11-1646; Lettre à Morus from 5-2-1649 (p. 1318-20). Certain formulas from this corpus, if they are isolated,

corresponds to a certain tradition of his reception, which is of the greatest of consequences from the point of view of history, not in regards to philosophical doctrines but the ideas which contributed to concept

---

can lead one to believe that men simply do not have a soul comparable to that of animals, what Descartes accepts calling a "bodily soul" (6ᵉ *Réponses*, p. 530, A. Morus, p. 1318), which is to say, that which corresponds to the functions of the body, this "animal machine", namely animated and living (*Traité de l'homme*, p. 873). And yet, we could say that it is this "bodily soul" (which is nothing more than the body envisioned from the viewpoint of its functions), which immediately and directly animates the living body, (animal or human) since (4ᵉ *Réponses*) it is not *immediately* our properly human soul ("spirit", "thinking thing", "reasonable soul", which man alone possesses), which moves the body: it only intervenes by the central demand of the "animal spirits" (whose flux functionally resembles what we today would call "nervous impulse", despite being produced within the boiler room of the heart, which sends them to the brain and from there to the nerves and muscles, declares *Discourse on Method* V), which effectively move the entire body in a profound unity of the organism (6ᵗʰ *Meditation*); and sometimes it doesn't even intervene at all, Descartes states. If we neglect this precision (that man can also be said to have a "bodily soul"), that 1) the animal body appears to be inanimate, non living, non animal, and that 2) man completely appears to be different than animal, even as a body. However, if we take into account the existence of this "bodily soul", then the Cartesian position can be presented as containing both *a certain resemblance* between animal and man (the same physiology and psychology can be applied to the study of the body and the soul which is attached to it in so much as the body is alive, "wherein we can say that animals without reason resemble us", as *Discourse on Method V*, p. 157 states, and this is indeed what the history of the sciences have shown, according to the comments by Simondon) and a radical difference in nature, since man is the lone possessor of this soul which Descartes calls *res cogitans*, thinking thing (which is so directly united and joined to the body and its functions that the entirety of existence finds itself affected by it).

formation in psychology and even the determination of its effective object. Descartes' doctrine, as it is discussed, can appear shocking if one worries about animals and fears that it will lead to their mistreatment, but for Simondon the most important question here is not discussing this problem, whether one should agree or disagree with it,[8] because its point of view is historical: it is this "Cartesianism", understood as such, that beyond all the reactions and sometimes passionate rejections of it, which "won" historically speaking and which, in contemporary psychological science, overturned and destroyed the Ancient conceptions at the same time it found itself overturning the Cartesian version of the cogito in order to distinguish in nature "the reasonable soul" and the "bodily soul". Such is Simondon's thesis on the "dialectic" of the whole that corresponds to the history it composed. Cartesianism, which wants us to be able, from a scientific point of view, to sufficiently recognize the animal in its behaviors, psyche, nature, in considering it as a machine, animated of course, but lacking rational thought (in the sense of a

---

8 Nor, for example, if Descartes refused the idea that animals possessed life, sensibility, and desire ("appetite"), which he expressly denied having supported in 6ᵉ *Réponses aux objections* (p 530) as well as in the letter to Morus from 2-5-1649 (p. 1320), where he merely says that, whatever we declare in terms of thought for animals, we can neither prove nor demonstrate they do not possess it, "because the human mind cannot penetrate their heart" (p. 1319), which Simondon quotes.

reflected thinking of the Cartesian *cogito*), not only corresponds, despite certain scandalizing protests, to what "animal" psychology initiated (ethology) starting in the 19th century, but above all what psychology in general ("human" psychology) has become, whether in the form of experimental psychology, behaviorism ("soulless psychology", according to Watson), or even in a more recent manner, in the form of "cybernetics" and cognitive science stemming from "Artificial Intelligence" (since 1946).[9] In its own way, Cartesianism, led to homogenizing, as scientific disciplines, animal and human psychology, in making psychology as well a part of biology, itself conceived, in terms of its principles, as a "machine" (if one takes this term in the true sense of how Descartes used it). Of course, to do this, one must put a parenthesis around what Descartes wanted to establish: the existence of the properly rational human soul (which, for its part, is not a possible object for empirical psychology but directly recognizes itself more easily than anything bodily). This effective historical "dialectic", that led to the current psychological sciences, in a sense was

---

9   We could compare the critique of Cartesianism with the critique Simondon makes, in *Du Mode d'existence des objets techniques* (Aubier-Montaigne, 1958, 1989), of N. Wiener's conception of cybernetics which relies upon "an abusive assimilation of the technical object with the natural object and more specifically with the living" (p. 48; see as well p. 110 – seq. and 149 – seq.).

understood in the objection by Gassendi: "as the soul of beasts is material, that of man can be as well".[10]

Animal and Man in Light of the Ontogenesis of the Vital and the Psychical

But, in conclusion, one is led to wonder what position Simondon himself holds. Indeed, it is not enough to look at a summary of the opinions to which Simondon accorded a propaedeutic value but in which he was not confident enough for the elaboration of thought, nor the consideration of what a history would present to us as a fact. It would still be necessary to verify, if the fact is established, what amount of rationality can be determined in the evolution that corresponds to it, to understand what it actually signifies and from what point of view. We propose, towards this end, to examine the manner in which its properly philosophical reflection has shown the necessity to pose the problem. Because not all questions are equivalent in philosophy. No question becomes philosophical

---

10 Fifth objection in *Méditations métaphysiques*, p. 471. We see how one can apply to the entire history of contemporary psychology what G. Canguilhem specifically said in regards the development of psychology as a science of internal meaning in the 18th century: "The entire history of this psychology can be written as a misinterpretation of Descartes' *Meditations*, without being responsibile" ("Qu'est ce que la psychologie?", p. 371, in *Etudes d'histoires et de philosophie des sciences*, Vrin 1970).

21

other than by its elaboration, which in general transforms the meaning of the initial inquiry.

However, in his major philosophical work, *L'individu et sa genèse physico-biologique,*[11] Simondon asks: "how do the psychical and the vital distinguish themselves from each other?" (p. 151); and not: how do man and animal distinguish themselves from one another? The answer to this latter question depends to a certain extent, of course, on the former one; but not in a direct manner: in forcing himself to answer the first question, Simondon feels compelled to also focus on the relationship between the human and animal (in a very marginal way), which reflects the fact that the two questions obviously have a strong link but also that the analyses by which he began to answer this fundamental question could have led to false ideas about the human and the animal. In fact, the note from page 152 begins as a correction: "Which does not mean that there are beings which are merely alive and others which are living and thinking: it is likely that animals sometimes find themselves in psychical situations, only these situations which lead to acts of thinking are less frequent in animals". Thus the distinction made between the notions of living

11 PUF, 1964, the first half of his principle thesis, whose second part was published under the title *L'individuation pyschique et collective* (Aubier 1969, Millon 1995).

individuals and living individuals having a mode of psychical existence does not correspond to that of animal and man.

Nevertheless it is true we could have expected to find, in this work, a determination of what the human and the animal are (and of their relation), in so far as his proposed general *intention* is to study "being according to its three levels: physical, vital, psychical and psycho-social", the determined problem being to "replace the individual in being according to (these) three levels" and the means to do so being "to study the forms, modes, and degrees of individuation in order to replace the individual in being according to (these) three levels" (p. 16). Nevertheless, what he takes "as the foundational areas such as matter, life, mind, society", are not *substances*, but "different regimes of individuation" (ibid.), and, at the end of the day, this doctrine "supposes a concatenation of physical reality going all the way to superior biological forms" (including man and his mode of being social), but "without establishing a distinction of classes and genre", even if it must be capable of recognizing that which, within experience, leads us to considering the relation of an "individual" to a "species", and of a species to a "genus" (p.139 and p. 243).[12]

---

12 In a way, genre and species do not exist. Only individuals exist; and furthermore, actually individuals do not fully exist either: all that exists is

There is not an essential difference between man and animal, because there is by principle no essential difference from the point of view of generalized ontogenesis according to Simondon's philosophy, this ontology that is both general and differentiated at the same time. It is an ontology of differences, of difference as relation. Everything is being, in such a way that one must take its singular nature into consideration at every turn. Every individual reality, even everything that is not individual (the pre-individual). It is because being is a relation. Every genuine relation has a "degree of being" (p.11). It is via its relation to the totality of being and the possible modes of being, that each thing is being (even it is not a "thing" in a substantial sense).

The note from page 152 does not say man and animal are identical but that we cannot denote an "essence

---

individuation (p. 197). "The individual is not a being but an action, and being is individual as an agent of this action of individuation by which its manifests itself and exists" (ibid.) This makes existence of living beings, as species, genre, or whatever type of ensemble, relying upon a "nature", lacks a sufficient founding objective: no classification, and by consequence, no hierarchy of the living is founded objectively (p. 163). The manner in which they can be regrouped should not only take into account their "natural" characteristics (anatomo-physiological) but the manner in which they effectively live in a group and how they themselves form a society (p. 164), the manner in which they individuate the groups they form, which is to say in effectively individuating (in a "transindividual manner") the groups where they individuate.

allowing the founding of an anthropology" in order to recognize the differences between them. Even when they are of the same breed, from the simplest to the "superior", animals can be quite different from each other. And this is no less true in regards to man, even if we are starting from the moment of ontogenesis (from the embryo, to the adult, to the final period of aging). There are without a doubt natural determinations which orient and limit the possibilities of individuation, whether they are psychical or vital ("animals are better equipped for living than thinking and men better equipped for thinking than living"), but the importance of the circumstances, the creation, and becoming they initiate should not be minimized. But, the circumstances should not be considered as liberating a piece of dormant potential, which up until that point had been asleep but which was nonetheless a *determining* factor (p. 153). It is in posing a new problem that circumstances can lead the living to a resolution, which takes the form of a new psychic and collective individuation.[13] Whether it is possible that animals "sometimes find

---

13 The psyche is not primarily a superior quality that certain living things possess. "The genuine psyche appears when the vital functions can no longer resolve the problems posed to the living" (p. 153); the regime of life slows down, becoming a problem for itself because the "overflowing" affectivity, "posing problems instead of resolving them" (p. 152), no longer has the regulating power of "resolving into a unity the duality of perception and action" (p.151).

themselves in a psychical situation" and that these situations can "lead to acts of thinking" (it is perhaps not completely an affirmation "that they think" or that they "have thought"), would "simply indicate that a threshold has been crossed". But "individuation does not obey a law of all or nothing: it can be carried out in a quantic manner, by sudden leaps" (p. 153). If "thinking" can have a meaning for an animal (we have no idea what it could signify for it, if not by way of conjecture, to the extent, as Descartes says, we cannot not know what it feels), nothing obligates us to consider that thinking would come to it as a complete mode of existence (corresponding to an essence) and entirely new for it, but would rather be a multitude of small differences in its mode of relation with itself and its environment, which would first of all be experienced by the animal as new *problems*. Simondon does not preoccupy himself with showing animals think, this would not have any meaning within the framework of his doctrine; but he shows that the general theoretical means at our disposal, outside of classical metaphysical or moral conceptions, from the perspective of generalized ontogenesis stemming from a reflection on the physical, biological, and psychological sciences, in order to imagine in general what the psyche and thinking are, cannot exclude the possibility of both of them residing in a being starting from the moment it is alive. What is an animal? What is man? What are their relations? We cannot answer these questions in a rigorous manner from the point of view of theoretical knowledge, to the extent that the terms in which this knowledge expresses itself are notions which have above all a metaphysical and moral signification. But we cannot know in advance the capabilites of a being, once we find ourselves dealing with a living body. Even if we can observe lines of strength and domination, we cannot limit what an already individuated living being can do nor what relation it can enter into, whether it is a relation with what is already inside it (pre-individual) or with something it is not (transindividual and interindividual). And perhaps here as well, one finds morality and metaphysics.

Jean-Yves Chateau

Two Lessons on
Animal and Man

First Lesson[1]

Today we will be studying the history of the notion of animal life within the area of psychology. It is, in effect, one of the sources for the formation of concepts between the natural sciences and humanities, which becomes manifest through the very long development of the notion of animal life. It is, via other forms, the problem of the relation between intelligence, habit, instinct, and life.

What is an instinctive behavior? What are the characteristics of animal behavior in contrast to properly human behaviors? What notion of the hierarchy of function has been manifested throughout time by various authors? In what manner could this hierarchy of functions have a heuristic value from Antiquity to today? This is what I will try to demonstrate to you in essentially two lectures, which will deal with the recapitulation of the different historical aspects of the development of this notion,

---

1 This text is the transcription of the recording of an introductory course presented at the University of Poitiers from 1963-1964. The following footnotes and titles have been added by the Editors. Since the publication of this course, several other works by Gilbert Simondon concerning the animal have been published in French. These courses, which he taught at the Sorbonne, concerning Instinct, Perception, Communication and Imagination are published with Editions de la Transparance.

in relation to the manner in which it is presented within our contemporary times: the problem of animal life and instinctive behaviors. Naturally, this undertaking will also elucidate the notion of animal psychology.

Antiquity

Throughout time, we can say that, in Antiquity, the first notion that emerged is neither that of instinct nor that of intelligence in opposition to instinct, but rather more generally that of human life, animal life, and plant life. What appears to be quite clear, or clear at least for the Presocratics is that the human soul – and this has really surprised the historians of thought – is not considered as different in nature from the animal soul or the vegetal soul. Everything that lives is provided with a vital principle, the great dividing line passes between the reign of the living and the non-living much more so than between plants, animals, and man. It is a relatively recent idea to contrast animal and human life, and to see human functions as fundamentally different from animal functions.

Pythagoras

For Pythagoras, the human soul, animal soul, and vegetal soul are considered to be of the same nature. It is the body and its functions which establish the differences between the various ways of living for a soul incarnated in a human body, the manner of living for a soul incarnated in a vegetal body, or a soul incarnated in an animal body. What emerges out of these first doctrines of the identity of souls and their community in nature is metempsychosis: the transmigration of souls. Metempsychosis is an ancient doctrine that supposes the soul is a living principle not attached to the individuality of one specific existence or another. An animal soul can serve to animate a human body, it can reincarnate itself in a human body, and a soul that has passed through a human body, after a human existence, can perfectly come back into existence in vegetal or animal form.[2] Diogenes Laërtius cites the phrase by Pythagoras, which, according to some, was meant to be ironic, who passing one day along the street saw and heard a puppy getting severely beaten. Pythagoras approached the tormentors and told them: "stop it, that is one of my old deceased friends who has been reincarnated as this beast." Diogenes Laërtius seems to assume, in retrospect, that Pythagoras' intention was ironic. But it is quite probable that via the legend, it is almost necessary to consider that, if Pythagoras could have said such a thing it is due to the fact there was a popular belief in metempsychosis

---

2 For example, Empedocles, *Katharmoi*, fr. 117: "I was in other times a boy and a girl, a bush and a bird, a silent fish in the sea...".

and that he had used the belief, in order to stop the agony of this animal. In any case, what is revealed by this story is the basis for a partially primitive belief in the transmigration of souls at the origin of our Western civilization, which implies that the soul is not a properly individual reality. The soul individualizes itself for a certain length of time under the guise of a determined existence, but before this existence, it has known other existences, and after this existence, it could experience more still.

One shouldn't neglect the heuristic contribution of such a doctrine or belief, because through this belief the possibility of the continuance of life becomes manifest, the reality of the passage of something else, which is more than the individual. Once the individual is dead, it is merely his body that decomposes and something else of him remains. Moreover, it is this idea of a durability of souls, of the virtual immortality of souls that will be taken back up by the spiritualistic doctrine of Christianity, but with an additional innovation that is obviously quite important: the individuality, the personality of the soul. Souls are immortal, but could we say that they only can be used once for a temporal existence? And, after that, they are fixed within their destiny? However, for the Greeks, the soul is in no way marked forever by an existence. After one existence, it can experience others: the soul is in a way reviviscent. It reincarnates itself, exists again in varied forms of different species and can pass from one living thing to another, this probably being itself the basis for the belief in different metamorphoses. Metamorphoses are changes in the form of a living being, which as the result of a curse or some fault, finds itself transformed by the gods or another power into a different species[3]. For example, a man can become a bird or he can become a sea monster or he can even become a river; a tearful woman can change into a tree or a fountain, These are metamorphoses which, in the end, are changes in species that concern individuality in a relative way, but which suppose there is above all an underlying vital but in a certain manner conscious principle that is conserved despite the morphological transformation of the individuality. I stated earlier that this primitive belief in metempsychosis and the possibility of metamorphosis, which is to say the changing of the form of existence while conserving a vital principle could be used to elaborate certain doctrines like the doctrine of the continuity of life and species change.

---

3 For example Daphne was transformed into a laurel tree when she was being pursued by Apollo; Aura was transformed into a stream of water by Zeus; Demeter bore bees from Melissa's dead corpse; The Heliades, the girls of Helios, were transformed into poplar trees on the banks of a river. See Grimal, *Dictionnaire de mythlogie grecque et romaine,* PUF, 1951.

We are going to soon discover in the doctrine of Plato there is a kind of transmutation, but a transmutation in reverse, a regressive transmutation that is the first known form in Western thinking on transmutation.

Anaxagoras

In staying with the Presocratics, at least with the authors who came before Plato, we find the doctrine of Anaxagoras, who affirms that there is a kind of identity in the nature of souls, but that there is, so to speak, differences of quantities, quantities of intelligence, quantities of reason (of *noûs*), the *noûs* of a plant being less strong, less detailed, and less powerful than that of an animal, the *noûs* of the animal itself being less strong, less detailed, and less powerful than that of man. These are not differences in nature, but differences in quantity, in the quantity of intelligence, in the quantity of reason found between beings.

Socrates

The first person to introduce an opposition within Antiquity between the vital principle of the vegetal, the animal, or man, thus the first who is in a certain sense responsible for traditional dualism, is Socrates. Socrates, in effect, distinguishes between intelligence and instinct, and opposes, to a certain extent, intelligence to instinct. He establishes, if we can call it that (we can effectively use the term in this case, but even if later on it was a kind of abuse to use the same word) a humanism, namely a doctrine according to which man is a reality that is not comparable whatsoever to any other found in nature. Between the nature studied by Anaxagoras and man which is studied by the Sophists and Socrates, there is no point of possible comparison and one would be led astray to give all of one's mind and strength to the study of nature. Socrates regretted dedicating his early years to studying the phenomena of nature with Presocratic physicists and Anaxagoras. He then discovered that the future of man and man's fundamental interest is not in the study of the constellations or natural phenomena, but on the contrary, in the study of himself. It is not about knowing things, the world, physical phenomena, but rather, as it is inscribed at the Temple of Delphi: *"gnôthi seauton"*, "know thyself". The Socratic lesson is of introspection and development by consciousness and the questioning of truths we ourselves possess as if we were full of truths. It is not nature that has a potential of truth to deliver, it is us who in ourselves possess this potential as humans, because we are exceptional beings, we have this burden of potential truths to bring to the light of day. And because of this, between animal instincts and human reason, between animal instincts and human intelligence, there is a difference of nature. By this, all of physics, which is to say the theory of the world and nature, finds itself rejected and dismissed.

Plato

And this leads to a theory which is not completely dualist, but which puts man before natural beings, a theory that is to a certain extent once again cosmogonical and cosmological, it is Plato's theory which in its own manner expresses the preeminence of man discovered by Socrates. In fact, it is through man that the animal is considered by Plato. And we find that human reality becomes the model for everything. In man, we find the image of the three kingdoms of nature. And we find this image in the form of three principles: *noûs* (reason), *thumos* (heart, élan), *épithumia* (desire). The preeminence of the *noûs* characterizes man; the predominance of the *thumos* (instinctive élan) characterizes the animal; and finally, *épithumia* characterizes the plant. If man were reduced to his viscera, if he was reduced to the organs existing between the diaphragm and the navel, he would be like a plant. He would be reduced to *to épithumétikon,* the "concupiscible" faculty, the "vegetative faculty" which knows only pleasure and pain, approval or disapproval, linked to needs or satisfaction. There exists need and it is the principle of pain, because lack is the principle of pain. When the need is satisfied, there exists contentment. The pleasure of contentment in opposition to the sorrow and pain of need, such are the two modalities of *to épithumétikon,* the faculty of *épithumia,* the faculty of concupiscence. As for *thumos,* it is characteristic of animals. Animals are courageous and instinctive. They have élan, an instinctive inclination, they tend to defend their progenitors, they tend to attack an assailant, they tend to a certain number of behaviors naturally because of *to épithumétikon.* A horse, a lion, can be courageous like a man. But what they do not have is *noûs,* namely, the rational faculty of organizing their behavior by knowledge, the faculty of acting because one knows why one acts. The animal does not know why he acts; he is brought to acting via an élan, by a kind of organic warmth that exists inside of it, by an instinctive élan. This makes it possible to envision different animals as sub-human, degradations of man. And Plato, in the *Timaeus,*[4] envisions a theory of the creation of animal species coming from man. At the source was man, which is the most perfect and which manifests in himself all the elements that later allowed to create by degradation (this is what I earlier called a reverse evolution) of the different species. For example, man has fingernails. But fingernails are of no use for man. They are a feeble armor; it is not extraordinarily powerful to have fingernails. But by progressive degradation, we see

4 Plato, *Timeaus,* 39e, 41b-43e. 76d-e, 90e-92c.

emerge little by little the role of the claw. First for men, then women are born and find a better uses of their fingernails. Then, we head towards the felines for which the use of claws is of an incontestable interest and for which the claw is much more developed and belongs to what we today call the body scheme, which is to say, they naturally know how to use it. The manner in which they leap is already correlated to the placement of the claws to grasp, to constrict their prey, to tear their prey apart. Consequently, the existence of certain anatomical details which in man appear as being mostly useless make sense in an organizing plan of the world from which all other species emerge directly from man, via simplification and degradation.

This idea from the *Timaeus*, which is in a sense monstrous, and in a sense genius, is the first theory of evolution in the Western world. Only, it's a reverse theory of evolution. Man is first amongst all other animals, and by simplification, by degradation, implies that the development of a certain aspect of the human body, such as claws replacing fingernails, one can obtain a certain animal adapted for a specific lifestyle. We are not talking about separating man from other animal species by a rising and progressive evolution but to show, on the contrary, how, from a simple human schema, simpler schemas can be drawn, which are those of animals. We can compare this to other reincarnation myths: souls drink the water of Lethe after having chosen a body,[5] a body they have chosen in function of their previous existences and merit, those who have risen to the most possible knowledge of truth and practice of meditation will not miss the chance of choosing the body of a philosopher; for others, they will end up with a particular animal existence. If Plato continued in this series of degradation, he could even say that one could reincarnate oneself in the form of a species of tree. But it seems the notion of metamorphosis linked to reviviscence in the vegetal form was spread in Greece by Eastern religious beliefs which were not that important in the time of Plato, at least in the area of philosophy; in the area of poetry, perhaps one saw a bit more. Indeed mythology contains stories of transformations into certain kinds of trees.

As a consequence, it is important to note that there is a notion of hierarchy in the work of Plato. In the *Timaeus*: everything is hierarchical, the three kingdoms are hierarchical, but they cannot be considered as strictly distinct from nature, but rather as levels. Nevertheless, the difference in levels in the end include differences of nature. In any case, we see this subside between the animal and the vegetal, it appears there is a resolution of continuity since it is

5 See, for example, Plato's *Republic*, X.

not stated that animals degrade into plants. This was the first part of the doctrines of Antiquity. We can call them in a certain way axiological and mythical doctrines.

Aristotle

And now, we have the second item of the doctrines of Antiquity, the first objective naturalist doctrine of observation which is that of Aristotle, regarding the relationship between the vegetal and animal and between animal and man. First, Aristotle did not scorn the consideration of vegetal existence. For him, the vegetal already contains a soul, manifests an existence of a soul, from a principle which is a vegetative principle, what Aristotle calls *to treptikon*, namely, that which relates to developmental functions and growth. *Trepein*, *treptikon*, comes from *trephô* to nourish, to thicken, and to make grow. The *treptikon* is what in the vegetal presides over functions of nutrition. This is very important and shows an extremely large deepening in the observation of Aristotle: the functioning of the vegetal is not merely to nourish itself. Notice how the hierarchical view of Plato is replaced by a view based on observation. A plant nourishes itself, which is to say, it assimilates, it grows. It assimilates itself in taking something from the soil, air, and light, in recuperating the necessary parts for the development and growth of the tissues of which it is constituted. It assimilates. This is nutrition. But this nutrition is not merely for itself. A plant reproduces. And nutrition is in function of reproduction. So, in *to treptikon*, by the fact of developing itself, the vegetative, is an effect of a nutrition and this nutrition is in view of generation (the final principle). The vegetal is finalized towards generation, towards production of itself. It's growth is a growth with a view towards generation. Thus there are plants like certain types of cacti (and many other plant types) which develop, getting bigger for several years, accumulate reserves, which then flower, bear fruit, and die. The finality of their development, their entire temporal history, converges towards this production of seeds. During several years they accumulate nutritional reserves and water in order to flower and bear fruit. Here you see the deployment of the idea of finality as being relatively important because you understand quite well that we can quite easily make animal, vegetal, and human life hierarchical simply based on the plant having the faculty of feeding itself. Earlier, in Plato, it was *to épithumétikon*. The Platonic *épithumétikon* is replaced by *to treptikon*: it is no longer a value judgment but a judgment of reality and the result of a study produced via experience. Plants grow, they assimilate, and they assimilate in such a fashion that these assimilations converge on the possibility of self-reproduction. There is thus a certain *logos*, a certain orientation finalized in the way a plant develops and constitutes itself. This is remarkably important, because here you have the replacement of relatively egocentric or at least anthropocentric value judgments from the first period of Antiquity, that I called mythological, by a judgment of reality, which is itself a result of observation, and thus much richer than a value judgment, since it includes a relationship between functions, to know the temporal relation of succession, but also organization, the functional continuity between different acts of nutrition and the act of generation at work in plants.

Furthermore, there is another aspect of the Aristotelian biology: the notion of identity or equivalence between animal, vegetal, and human functions. While the same functions can be filled in these kingdoms by processes with relatively different operatory modes, it does not prevent them from being comparable. Here Aristotle introduces a new abstraction, by means of the notion of function, which is much greater than that of his predecessors. In animals, in addition to the *treptikon*, this faculty of growth, there exists *to aisthètikon*, the faculty of feeling. In the same way that *to treptikon* is made of nutrition and generation, the *aisthètikon* also combines two functions: first *aisthèsis*, the faculty of experiencing, of feeling, and *orexis*, the faculty of desire, which is the consequence of *aisthèsis,* characteristic of the animal. The animal is endowed with sensitivity and motor skills, motor skills in the form of desire, of élan. It is a bit comparable to the *thumos* that we found earlier in Plato's doctrine. Sensation is *hèdu kai lupéron. Aisthèsis* is the faculty of experiencing *hèdu kai lupéron,* the pleasant and the painful; the two qualities are *hèdu kai lupéron.* In fact to experience the pleasant and the unpleasant results in *orexis.* The élan that strives to avoid pain and searches for pleasure is the motor of every living thing, every living animal, because it is not clear that the plant experiences pleasure and pain. At the level of *aisthèsis,* there also exists a *phantasia aisthètikè,* a sensory imagination, a sensitive imagination. Finally, in the animal, at least in certain animals high enough up in the sequence of the living, there exists a simple memory, *mnèmè,* in contrast to *anamnèsis. Anamnèsis* is reserved to man because it supposes recollection, consciousness, an effort towards recall. The *mnèmè* is direct memory, spontaneous memory. And *anamnesis* is the faculty of memorization or recollection. There is thus in animals, at least in the most developed animals, sensation, sensory imagination, passive memory, desire and, as a result of desire, movement. What is missing in animals so as not to be like man? The animal lacks the faculty of reason, *to logistikon,* the logical faculty. The animal also lacks the faculty of free choice, *bouleutikon*, the free deliberation or more precisely the choice after the examination of all possibilities of action, free choice called *proairésis,* the preference given to what is logically preferable. Reason and choice are thus characteristic of the human species, but this human species is not strictly different in nature from animal species.

What is fundamental in the doctrine that I just presented is that it does not strive to provide mythological conceptions and above all morality at any level but, on the contrary, tries to show how the different vital functions express themselves in the plant, animal, and human. This aspect of continuity is particularly manifest in the notion of the imperceptible passage from plant to animal. Starting from marine animals or aquatic plant life, Aristotle reasons that one could also call trees "land-oysters". The manner in which oysters develop in the sea is not essentially different than how plants develop on land. In fact, oysters are fixed and develop and grow progressively via the accumulation of matter they construct, and grow their shell by adding successive pieces that remain marked afterwards, to such an extent that one can see the growth marks of an oyster shell almost in the same way one can see how old a tree is when one cuts it down and counts the rings. Many sea animals that mature in shells, indeed grow like a tree thickens its trunk in adding the successive generating rings of wood. And for this reason alone, at the lowest level, it is impossible to state whether we are dealing with a plant or an animal. Thus, one shouldn't be caught up in being hierarchical at all costs. There exists, so to speak, a common trunk in both the plant and animal kingdoms. And this still remains today. We call protists living beings that we cannot clearly distinguish in a certain way from amongst the animals or plants. Protists would be the living beings anterior to any possible differentiation in animal or vegetal.

Analogy, moreover, functional analogy goes even farther and it is starting from this analogy we are able to think with a certain depth, in the work of Aristotle, about the instinct. For Aristotle, the ways in which bees construct their hive in order to shelter their honey and youth, is parallel with the method in plants that produce leaves in order to surround and protect their fruits. Instinctive dispositions in animals like the construction of a hive, the construction of a nest, are comparable to certain modes of growth, which, have a visible finality in plants. What animals do by various movements such as the way bees construct their hive and benefit from the honey comb inside, is the construction of a structure comparable to what we see develop in the growth of a plant, a process with a view towards generation, reproduction. Only, their operatory modes are different. The animal and vegetal world are different, but there is a functional identity, so to speak, a functional parallelism, between these distinct operatory modes. In the less developed, least differentiated animals, the functions that liberate and define other higher animal forms such as imagination, anticipation, this *phantasia aisthètikè*, already indicates a certain experience and allows the use of the experience in similar cases than those experienced. The *phantasia aisthètikè* does not exist in ants, worms, or bees, states Aristotle. Ants, worms, and bees have no imagination whatsoever. They work and construct like a plant grows. The society of ants or the society of bees constructs its hive like a plant grows and constructs its branches and leaves during its development. This is where instinct appears. Instinct is a certain faculty of constructing as if it were a way of developing, like that of a plant. What is instinct in animals is, in plants, the fact of growing in such and such a way, of developing a certain foliar scheme, formula, of the given vegetal form, with very specific characteristics. Consequently, instinct, in as much as it is an operatory mode of construction of a hive or an anthill, instinct is equivalent to a structure of development. It is specific. Instinct is part of specificity, it is a drive in animals and more specifically social animals, which is equivalent to growth defined by specific lines in the plant.

In the most differentiated, most developed animals, there exists, on the contrary, not only this *phantasia aisthètikè,* but a certain habit, a habit which enables animals to learn, and by the acquisition of experience, they acquire a certain capacity to foresee what is presented and to palliate the different inconveniences of possible events. This imitates human prudence, namely prediction, *prudentia* being the faculty of foreseeing and adapting ones behavior to the events that unfold. Habit in animals is a kind of experience that imitates human prudence. Imitate here means that which is a functional analogy to prudence, but with different operatory modes. As with the way plant development imitates that of ants and how bees construct their hive, so it is that habit in animals imitates human prudence. Human prudence can use reason, it can make use of *bouleutikon,* of *logistikon,* of *proairésis.* Animals cannot make use of *bouleutikon,* of *logistikon,* of *proairésis,* but, despite this, habit imitates this prudence, a prudence which appeals to reason, free choice, and calculations of chance.

Thus, even if we admit it – and we have to admit it, that according to Aristotle reason is properly human and specifically characteristic of man, there exist continuities and functional equivalents within the various levels of organization between the different modes of living beings. Aristotle's oeuvre is essentially a work of biology and natural history: you can see to what extent Aristotle went in developing the notion of function, in flushing out the different vital drives of the notion of function, which allow us to align parallels between beings whose mode of existence and structure are very different, but from the point of view of life, are conceived as a chain of functioning which is nonetheless comparable. A general knowledge of living beings becomes possible through Aristotle's notion of function, and even psychical functions that one can more or less discover through observation or introspection in analyzing man, can correspond to the functioning of other living beings. At the heart of this doctrine itself is the notion of function which allows that of the notion of equivalence to be implemented, an equivalence which can go from the vegetal to the animal and the animal to man, and even from man to the vegetal, because what counts are the functions and not simply the species. There can be extremely different degrees of organization, this is not important. It is still however possible to equate the functional realities of one species to another. And it is here that we can see biological science in the work of Aristotle. There is biological science, because there is a "great hypothesis". In Aristotle, it is called *theoria*, a theory.

It is the theory of functions. It is the theory according to which all species live in the same manner. Or one could say: all species *live*. And thought, reasoning, *bouleutikon*, *logistikon*, and *proairésis,* what appears as a specific characteristic of one species is perhaps indeed characteristic, because it doesn't exist in another species, but the functions which are filled by the characteristic gifts of a species are not unique to the species. The means that a species has of answering to its needs are unique to it. The specificity consists in the certain faculties that the species possesses and the others do not. But, furthermore, the reasons these faculties are implemented and the functions they serve have nothing specific about them at all: life is the same everywhere. In an oyster, in a tree, in an animal, or in a man, life has the same demands. For example, in growth and reproduction, we find the same demands corresponding to the same parallel functioning. They can be achieved with extremely different operatory possibilities. What man does using *bouleutikon* or *proairésis* or *logistikon*, an animal will do out of habit if it is sufficiently reared, or simply in the way it constructs a hive or anthill if it is not endowed with greater abilities. What is not possible with certain faculties can be achieved by others and the functions remain. The means change according to species but the functions remain. And this is perhaps what is most profound, this is what truly is the grounding for a theory of life in the work of Aristotle, the theory being: there is an invariant, and this invariant is life; the functions of life; the means used to fulfill these functions change with species, but the functions remain, life is an invariant. And here, you can see in this, Aristotle established a science. He is indeed the father of biology, and he included psychology in biology because psychical functions like reasoning, deliberation, and free choice are all part of accomplishing operations that are part of life, operations that have a signification in vital functioning are comparable to other vital functions accomplished by other means. One could say that man thinks, and that, in thinking, in using his rational faculties, in using *bouleutikon*, *logistikon*, *proairésis*, he does something that the plant does in developing its leaves, giving birth in a certain way to its seeds. Thus there is a continuity of life and permanence of life from one species to another.

The Stoics

After these discoveries, which could pass for the foundations of science within Aristotle's doctrine, the Stoics, at the end of Antiquity, return in a certain way to the ethical doctrines of Aristotle's predecessors, the Platonic or Socratic doctrines. The Stoics, in effect, deny intelligence to animals and develop the theory of instinctive animal activity. They contrast the human functions of liberty, rational choice, rationality, knowledge, and wisdom with animal characteristics that come by instinct. It is the Stoics who develop the most complete theory of instinct. And one can call them the founders of the notion of instinct for ethical motives. They want to show that the human is a being apart from the rest of nature. That all of nature is made for man, that he is, so to speak, the prince of nature, that all converges around him, that he is the king of creation and that, consequently, he is endowed with functions which are not found in any other living beings. Note well that this comparison (between man and animals), contrasting instinct and reason, is twofold: for certain Stoics, it merges with the theme of morality, a quite easy amplification of the theme of a thinking reed. Man appears inferior to animals in regards to everything having to do with nature, and instinct, but he is incomparably superior to them in everything having to do with reason. Thus if you take certain passages from Seneca, you will find numerous elements in Latin Stoicism and rich comparisons between living beings who are living animal beings and perfectly adapted to their function by nature, and man who is, as it were, from the beginning, maladapted. For example, Seneca states that one finds in all living beings natural defenses. Some have beautiful fur that protects them from cold, others have scales, others have quills, others have a slimy skin that makes it hard to grasp them, others are enveloped in a hard shell. As for man, he has nothing. When he is born, he is *dejectus*, he is placed on the ground, he is incapable of moving, while small birds are already capable of finding their food, while insects are born knowing how to take flight. Man knows how to do nothing. He is, as it were, disgraced by nature. He must learn everything and he must for many long years depend on his parents in order to be able to earn his life and guard against the principle dangers that lurk. But, in contrast, he has reason. He is the lone of all animals to stand up straight, to gaze, to have his eyes towards the heavens. There is an amplification which is an oratory amplification, to a certain degree, but which is nourished from the idea of a disjunction between man and nature. This basic disjunction between man and animal, it would seem, has as its principle initiatory aspects, certain doctrines, perhaps Orphic doctrines, Pythagorean doctrines, or came from Orphism or Pythagoreanism, that showed man had a destiny apart: all the rest of creation, it is the world, it is nature, it is limited to itself, but man is of another nature and he will discover his true destiny in another world. Perhaps in the Stoics there is the beginning of this quite vast aspiration of escaping the world, which manifested itself at the end of Antiquity; in any case, the idea was that nature was insufficient, nature as such was lacking and the human order is of a different order. They were the founders of the notion of instinct in order to show there is an enormous difference between the principle of animal actions and the principle of human actions. The goal is ethics.

Conclusion of the First Lesson

We will thus distinguish, to summarize, within this period of Antiquity, the Pre-Platonic or Platonic doctrines that are essentially of an ethical nature; then the relative doctrines of Aristotle or which were developed around Aristotle (like that which we find in the work of Theophrastus for example), are above all doctrines of functional correlation between principle psychical activities and the different activities existing in animals and even plants, corresponding to a certain degree, with a naturalist theory of psychical functions; then, finally, the third point, a return to the ethical doctrines with the Stoics, thanks to the notion of instinct, essentially comprised of automatism. The animal acts by instinct. What the animal does that resembles man, it does by instinct. Whatever this may be, man does it by reason. Consequently, man is of a different nature than animals and plants.

Second Lesson

Problems and Challenges

We ended our study last time saying that at the end of Antiquity, the Stoics deny intelligence to animals and develop the theory of instinctive activity, namely an activity comparable to intelligence in its results but in no way based on the same internal functions. Specifically, animals are not as attached to the cosmic fire as man, to the *pûr technikon*, to this artisan fire which cuts through all things, assembles them and gives them a meaning. But above all, despite everything, Antiquity constituted an opposition, and crystalized an opposition between theories that are fundamentally naturalistic, physiological and those by contrast that would tend to consider man as a being separated from the universe. Nevertheless, despite this opposition between natural behavior and human reason, behavior by beings above all made of matter, what we generally find in Antiquity is the notion of gradation between animal reality and human reality, either via ascending gradations as we find in the work of the physiologists or via degradation as we find in the Platonic doctrine. But whether we are dealing with gradations or degradations or whether it is the distance we admit exists between the animal reality and the human reality, we are nonetheless led to indicate a progressive phase of a possible continuity. Whether it is a degradation which goes from man to animal or whether there is a gradation from the simplest of animals such as fish born at sea, out of water, and passes progressively towards man via an ascending series, this supposes, whatever the distance there is between human reality and animal reality, in the end, deep down, there are fundamental functions, behaviors, attitudes, and mental content of the same nature in both man and animal. This measured continuity, this functional equivalence, we saw it presented in the most clear, sensible, detailed fashion, and finally as the closest thing in Aristotle's teaching to a scientific theory. And Antiquity remains, around Aristotle's teaching, a vision of the relationship between animal reality and human reality which is an intelligent, generous, non-systematic, at least from the outset was non-systematic, non-dichotomous, in its results if not in principle, and as a consequence authorizes parallels, comparisons, prioritization, but not a prioritization towards ends which are normative oppositions between one natural reality and another natural reality. What comes out of the teachings of Antiquity is that what occurs in man and what occurs in animals is comparable. Comparable. Not identical but comparable: it is with the same mental categories, the same regulating concepts, and the same schemas that we can further our understanding of human and animal life, inside the general teachings of existence, of our relationship with the world, reincarnation, palingenesis, or the gradation and degradation of existence.

On the contrary, and we will try to see this today, the intervention of the doctrine of spiritual activity, starting with Christianity, but much more still at the interior of Cartesianism, constitutes a dichotomous opposition, an opposition that affirms two distinct natures and not merely two levels, putting on one side an animal reality devoid of reason, perhaps even of consciousness, and most certainly some sort of interiority and on the other side a human reality, capable of self-awareness, capable of moral feelings, capable of being aware of ones acts and their value. In this way, we can see, and this is really important, that the most systematic teachings are not, as we could say, the teachings of Antiquity, but, on the contrary, those of a certain number of priests of the Church, reflecting moreover with moderation the work of St. Thomas that partially goes back to Aristotle, and who is one of the most moderate of all Medieval authors and above all, in the end, the Cartesian teachings which are quite frankly totally systematic and dichotomous doctrines.

Let's have a look at the first doctrines, namely those doctrines which were above all ethical in nature, the metaphysical doctrines of religious inspiration and ethics. Then, afterwards, we will look at the Cartesian system with regards to the notion of animal life that presents a trait-by-trait contrast of human life and animal life. I will allow myself to say that this precisely excessive, bizarre, scandalous character of the kind found in Descartes' doctrines provoked a movement of thought that, in final analysis, was perhaps favorable to the discovery of the scientific theory of instinct, of behaviors that are animal behaviors, and finally by a very curious turn of events, to a contemporary theory of human instincts. That is to say, there is finally a dialectical movement which was produced by the research and comparison of human and animal life: at the point of departure, in the Ancients we had a kind of phenomenological aim, that starting with the principle aspects of human and animal life, prioritizes human life in relation to animal life but it does so without a rigorous or passionate opposition. Then we see the birth of dualism, that uses the animal as a kind of foil for man, that treats the animal as non-human, that makes the animal a being of reason, namely a fictive being, a living or pseudo-living being that is precisely what man is not, a kind of duplicate to an ideally constituted human reality. And finally, in returning to things often produced when theory encounters reality testing, notions such as the animal are found to be generalized and universalized enough to permit the thinking of human behavior itself. This is characteristic of the development of the problem of the relation between human and animal life during the 19th and 20th centuries that denies Cartesianism not in order to state the animal is a being of reason and has an interiority, a being that has an affectivity, a being that is still aware, and thus has a soul, which would simply be the reversal of Cartesianism, but which reverses Cartesianism in a most unexpected and singular manner: the content of reality you put into the notion of animality, this content allows us to characterize man. Namely, it is by the universalization of the animal that human reality is dealt with. Here there is an evolution of a scientific theory that is most certainly of a dialectical kind. From Aristotle to Descartes, from Descartes to contemporary notions of instinct, biological notions of instinct, there is truly a relationship between thesis, antithesis, and synthesis: Cartesianism constituting the antithesis of the theory of Antiquity, according to which human reality and animal reality are in continuity. Descartes affirms they are not in continuity. Finally, the contemporary thesis once again reaffirms they are in continuity, not merely by the reversal of Cartesianism, but in saying what is true about the animal and what is true about man.

While the Ancients strived to say: what is true of man is true to a certain extent in the animal, above all to the extent that he is a superior animal (this is Plato's theory of degradation); afterward, Cartesianism says: what is true of man is not at all true in regards to the animal. The animal is part of the *res extensa*, man is part of the *res cogitans*, is defined by *res cogitans*; In the end, contemporary theses consist of saying: what we discover at the level of instinctive life, maturation, behavioral development in animal reality, allows us also to think in terms of human reality, up to and including social reality which in part is made up of animal groupings and allows us to think about certain types of relations, such as the relation of ascendancy-superiority, in the human species. There has been a dialectical movement here we have been striving to trace.

The Apologists

Let's begin by looking at the first authors who tried to define a relatively dualist relationship between human reality and animal reality in the work of the Ancients, or more precisely after the period of the classical antic world, in this period that initiates the theory of action as being prior to knowledge. For example, we find amongst a certain number of Apologists like Tatian, Arnobius, and Lactantius[6] an attitude

---

6 Tatian, Christian apologist, then gnostic, born in Assyria

of extremely powerful ethical dualism that does not strictly speaking have its sights set on contrasting man to animals but indeed Christians to the entirety of non-Christians and animals. In the end, reason, this faculty that was exalted by the Ancients, is humiliated by the saying that the Christian alone differs from animals and all other men are not different from animals. You can see the responsibility of the ethical reality that is found incorporated in this doctrine. You don't have to be moved by this, you know one of the first councils thought women did not have souls, for reasons which were perhaps the same as the one mentioned here: don't look at this as merely a bad joke, but in a general sort of way, we always end up, where one has to prove one's own interiority, believing that one has a soul, that one thinks oneself (*cogito ergo sum*). But other people, viewed from the outside, are little by little repudiated to the point of their whole nature. Barbarians, or rather those beings sexual dimorphism separates to a certain extent from this experience one has of one's own interiority, one can indeed suppose they are merely products of nature. This is because the notion

---

between 110 and 120. Wrote a *Discourse on the Greeks*. Arnobius, Latin author, born in Africa, contemporary of Diocletian, died in 327. He taught rhetoric in Numidia and had as a student, Lactantius. Lactantius, Christian apologist, died around 325. Education in Africa. He mentions Tertullian and Cyprian.

of the soul is linked directly enough to the experience of interiority, the experience of consciousness, to the exercising of consciousness. As soon as there is an ethnic, cultural, sexual difference or any other sort of species, it can be sufficient enough to constitute a barrier for the attribution of the soul to be refused because the others will be not be experienced as being very similar to the subject that actively experiences its own interiority.

Saint Augustine

Saint Augustine, who is linked in close proximity to the Antic culture, on the contrary thought that animals have sensitive souls. He thought animals had needs, that they suffered, he knows they struggle with pain, he knows they struggle to maintain the integrity of their organism. Saint Augustine also thought, with the support of experienced observation, that animals remember, imagine, and dream. One can look at a dog sleeping and see it thinking it has caught some sort of prey, and even bark, and suddenly take on the gestures of grabbing the prey within its teeth, opening and closing its mouth as if he had grasped it. This is basically the external manifestation of a dog dreaming by way of explicit attitudes. Despite it all, Saint Augustine thought everything is instinctive in animals, that the different abilities and constructions are explained by the senses, imagination, and memory without the intervention of the soul, at least in regards to a reasonable soul, the soul such as the human soul, endowed with a moral sense and the exercise of reason.

Saint Thomas

The scholastics, who themselves are animated by the memory of Antiquity and specifically the memory of Aristotelian Antiquity deny reasoning in animals. But with Saint Thomas, they recognize and even make explicit that animals do have intentions, distant ends for which they work, and which are consciously perceived by them. Thus the swallow that collects a piece of mud to construct its nest does not do so out of pleasure. It accumulates mud because it needs the mud to construct its nest and it has the intention (namely the interior experience of finality) of constructing the nest. Intention is the fact of literally being "turned towards", having the activity oriented towards the realization of an end. Thus the swallow has the intention of constructing the nest, it is the distant end of its activity, one should not say that it acts out of pleasure, because the mud pleases it. This distant end is perceived, according to Saint Thomas, by *aestimatio*, namely by a relatively qualitative impression and not reflexive nor rational, but despite this, it is a representation. It is not totally logical, absolutely schematic and structured, but it is indeed a representation. Man possesses a logical and rational faculty for thinking allowing him to conceive of ends with much greater organization and clarity than that which allows the swallow to have *aestimatio* of constructing the nest. Nevertheless, for Saint Thomas, the finality of animal behavior corresponds to a certain representation. We can see here how Saint Thomas takes up, while developing it a bit more, Medieval conceptualism, the Aristotelian doctrine (the doctrine of finality and this doctrine that prioritizes activities in animals). But, if a certain moderation (let's call it phenomenological and scientific) was conserved by Medieval authors next to a type of dualist passion, above all manifest in the work of the Apologists (which made a myth out of the animal, the myth of that which is not a being of faith, the creature that does not have a direct recognition of God), next to that, there is during this first period, the memory of Antiquity.

Giordano Bruno

However, the Renaissance intervenes as a very rigorous rediscovery of the relation between the animal and human psyche. We could even say that the Renaissance exalts the animal psyche to avenge the dualism of the Apologists, putting the animal psyche above the human psyche in order to teach us lessons. Here as well, there is a certain theory, a certain passionate aspect of the animal which mythologizes it: the animal is thus nature, the *phusis* teaches man, that teaches him lessons, either about purity, or devotion, or ability, or even about intelligence relevant to a discovery of a goal. The reversal of the Renaissance occurs via an inspiration that is extremely close to that of the élans toward the *cosmos* of the ancient Platonists in the work of Giordano Bruno. Giordano Bruno, burned at the stake in 1600, is one of the most powerful philosophers of the Renaissance. He is a metaphysician of the vastest of thought, the most vigorous of scientists within the generality and span of his doctrines. He concluded with a doctrine according to which an innumerable amount of different worlds exist, other inhabited earths, not merely our own, but other inhabited planets in which life also developed. According to his doctrine, animation, which is to say life, is not merely a fact for beings at the scale of life as we know it, but it can also be a fact for stars (there are animated stars), life can exist in elements where we don't believe it to exist. Even the stone in its own way experiences certain affections. Life and consciousness are not phenomena that only appear with forms like that of human forms; life and consciousness begin existing at a cosmic level. Giordano Bruno's theory is a cosmic theory. To this extent, it is certain that animals are considered as beings, agents of a universal force and consequently should not be held in contempt, they should not be considered as inferior beings or caricatures of man. We can perhaps see a relationship to this type of thinking with other various movements in thought like those developed in Italy by the likes of Saint Francis of Assisi and his way of considering animal reality.

Saint Francis of Assisi

For Saint Francis of Assisi, animal reality is not at all something vulgar and sordid. It is part of the universal order. Animals, in their own way, recognize the glory of the Creator and the harmony of Creation and, in their own way, adore and honor God. This is why it is not impossible, if one attains the right level of purity, of moral purity and simplification of oneself, to directly be understood by animals. Communication between man and animal was only rendered impossible due to human sin, by the thickening of conscience, the vulgarity and heaviness of habit; but a man who has purified himself enough, who is sufficiently inspired, who is conscience of the Universe and Creation, and who loves God, can be understood by animals. You have heard about the animals gathering to listen to Saint Francis of Assisi? What's even more interesting are the legends developed during this time period indicating the possibility of granting the notion of saintliness to animals. The notion of saintliness in religious and ethical thinking was not merely reserved for human beings but there also existed animal saintliness. This is a thinking that goes well with certain conceptions of the Renaissance. The Renaissance discovered a relationship between man and things, between man and the Universe. Instead of considering human reality as a special creation by God for which the rest of the Universal order was finalized and to which it is subordinate in an absolute manner, it is actually rather according to an aesthetic order that the relationship of the human to the animal is thought. The entirety of Creation is harmonious; the place of man is complementary to plants and animals. There is a universal totality. It's the notion of the Great Being, this kind of pantheism that developed to a certain extent during the Renaissance; in the Christian authors, it is not a pantheism of course, and it becomes a theory of the harmony of the Universe, the universe as God's creation; but in the work of the pantheistic and naturalist writers, there was truly a renewal of ancient pantheism.

Montaigne

The echo of the doctrines of the Renaissance is found in the work of those authors who directly prepared the way for Cartesian thought but who in no way accepted the dualism between man and animal. For example, the case of Montaigne; Montaigne represents more the state of mind of the Renaissance than that of Cartesianism. He is fundamentally monist, which is to say all psychical faculties existing in animals are the same as those existing in man. For Montaigne, animals judge, compare, reason, and act the same way as man; the same way and even better. You know, Montaigne has a kind of undulating thought; it is difficult to grasp exactly what one could call a system from his thinking. It is much easier to grasp his intentions rather than his system. Montaigne's intentions are quite clear: like the Apologists, he has the intention of humiliating pure reason, that which produces systems, but even more than reason, he wants to humiliate human pride, because the human pride for theories too systematic in nature is what leads to us burning men, to religious wars, it's what leads to the most bitter and destructive conflicts for mankind. Thus one has to reintegrate man into the order of Creation, make it so that he conceives of himself as being a close relative to animals who live in an ordered manner, who live much more directly linked to natural processes. This is why Montaigne evokes the goats of Candie whom, once they have been wounded (by an arrow), by the lone instruction and mastery of Nature, go search out the specific plant, the herb Dittany, and eat it in order to heal themselves. Instead of giving us something to be prideful about by saying animals act by nature and that humans, when sick, choose such or such medicine by use of reason, we would be better to consider that animals have the honor of having nature as a "school teacher".

And yet, you see that despite everything, there is a shift in meaning. This is what is important in Montaigne's theory (this is taken from "The apology of Raymond de Sebonde").[7] The doctrine that intervenes here is subjected to a shift in meaning because, as you can see, Montaigne perfectly distinguishes what a man acquires from trial and error by a relatively delicate use of reason, a reason that can integrate experience, a reason that can be subjected to error, and it is a reason that is never completely immediate, while the goats of Candie, once they are struck with a malady, directly go to eat the so-called plant Dittany. Here there are obviously two different types of behavior, and Montaigne knows this very well, since he states that animals are indeed lucky to have the honor of such a certain teacher, nature, of acting according to a behavior that is different than that of humans, rational behavior being merely one kind of existing behavior. While Montaigne shows that it is the animals who are superior since they do not even have to pose the question of knowing which medicine to

---

7 Montaigne, *Essays*, Book II, Chapter XII.

choose, they know the medicines directly, they have a "certain teacher", they don't make mistakes.

This is the open door to dualism: under the idea of exalting animals and showing that man does not need to be so prideful of his humanity, because in the end, he is not really superior to animals and perhaps it is the opposite, since man makes mistakes, since he is obligated to have recourse to reason while animals do not even need this reason, shows their superiority. They are more directly in relation with nature, when one says this, one is implicitly admitting that the rational process, which is to say the process of apprenticeship, is different than the instinctual or instinctive process of animals, that is more immediate, more direct behavior. And it is indeed starting from this opposition that there will be a complete distinction between on the one hand the inspiration of the Renaissance, which is a naturalist, monist inspiration, and on the other hand Descartes' system, which is a dualist system, more dualist than any other dualism since Antiquity, more dualist perhaps than the Apologists like Tatian, Arnobis, and Lactantius in declaring the Christian is completely different than other men and animals.

Descartes

Indeed, according to Descartes, animals possess neither intelligence nor instinct. The animal is a machine, an automaton. What we have up until here explained as instinct, by a psychical analogy of intelligence, but a more compact, more concrete, less conscience analogy, more enveloped, is explained by automatism, but careful, this could be an instinctive or physical automatism. The Cartesian doctrine is that of a physical automatism, namely an automatism of beings, bodies, attitudes, and movements, without soul and instinct. One must understand that instinct within a doctrine like that of Montaigne is not reason, but psychological. It is a reality that is of the psychological order. This can be said for the Stoics as well. Descartes is the first who said animal behaviors are not instinctive. They are not instinctive behaviors: they are mechanical. It is not at all the same thing, because this could easily raise some confusion: one can say that what characterizes instinctive behaviors (moreover this is false, we will see why in a second is automatism). This has often been said since the Stoics. But what we want to talk about is psychical automatism, an automatism comparable to that which we obtain or think we obtain once we, for example, undertake a very thorough apprenticeship, an apprenticeship where we learn everything by heart, and can fire off a series of numbers or words or a text without thinking about it while doing something else at the same time. These kinds of activities, once they are established, can take place starting with a triggering process as an initial stimulation, such as the recitation of a text that one must start from the beginning in order to dictate the entire text completely, establishing an automatism that one could call an automatism of a psychical nature. But this is not at all the type of automatism Descartes is talking about. He describes an automatism which is far from being analogous to intelligence, or acquired habit, and learned. His is an automatism of matter, of the *res extensa*, namely something comparable to the functioning of a machine, due to the form of its pieces. When a spider constructs its web, it acts precisely like a weaving machine (a loom). When a mole digs its molehill, it acts like a shovel, namely as a tool made to disperse with the dirt in a specific manner. Animals are conformed to a certain type of action that is moreover generally quite narrow. Outside of a specific material manipulation corresponding to their bodily conformation, they are extremely awkward, and incapable of solving a true problem. Far from the industriousness of animals used to show the superiority of animals, these wonderful examples actually go against showing a kind of instinct in animals if we want to consider instinct as something psychical. There is no such thing as instinct. There is merely bodily automatism. This is what Descartes says: "even though there are certain animals who testify to being far more industrious than man in some of their actions, we can see that these same animals don't appear to be more so in other actions: the manner in which certain animals prove to be better than us does not establish they have intelligence, because in this case, they would be much more intelligent than us and would be better in all things, but rather it establishes that animals have no intelligence at all, and it is nature which acts in them,"[8] which is to say the conformation of their body. They act via figure and movement. And in the same way one can do little else with a shovel besides shoveling, or use a loom for anything else but weaving, a spider is incapable of doing anything else besides weaving its web or a mole to shovel dirt and make its molehill from it. The animal, by its bodily structure, finds itself eminently apt to the functioning of its body, and outside the functioning of its body, it can do nothing. Of course, Descartes says, the human mind cannot penetrate into the heart of animals to know what is actually taking place (*Letter to Morus*).[9] But in the end, Descartes affirms that thought is enclosed within the feeling we have of it, that thought is thus conscious, he also affirms this: "after the error of

---

8 Descartes, *Discourse on Method*, Part V

9 Descartes, *Letter to Morus*, Februrary 5 1649

those who deny God, there is none who makes weak minds stray farther from the straight path of virtue than imagining that the animal soul is of the same nature as our own."[10] Which means the human soul is *res cogitans*, and all of animal reality is *res extensa*, without consciousness, without interiority. You will indeed notice the criteria Descartes uses in order to distinguish human reality and animal reality is this: human reality is distinct from animal reality because animals, like tools, can do one thing very well, and outside of that, nothing. No plasticity whatsoever. While the human being can place all difficulties in the form of a problem and progressively resolve them a step at a time, etc., basically the Cartesian method. This shows that man is not adapted to any one specific figure and movement (he does not have the conformation of the mole or the weaving ability of the spider), but because of reason, mind, of what Descartes calls, "having spirit", having a soul, having a rational faculty, by the fact man has wit, he can attack all difficulties and strive to overcome them by means which are progressive. Thus there is the negation of consciousness in animals, above all the negation of the faculty of rational acquisition, intelligent apprenticeship, and intelligent problem resolution. And we have the notion of automatism

---

10 Descartes, *Discourse on Method*, Part V.

in animal behavior and the suppression of the idea animal instinct.

Malebranche

Amongst those who adopted the Cartesian doctrine, none is more fervent than Malebranche. He has a wonderful argument for explaining why animals most certainly cannot have souls and that they do not suffer. He writes: "Animals eat without pleasure, they cry without pain, they grow without knowing it: they desire nothing, they fear nothing, they know nothing: and if they act in a manner demonstrating intelligence, it is because God made them to protect themselves, he formed their bodies in such a manner they mechanically and fearlessly avoid everything capable of destroying them".[11] This is taken from *The Search after Truth*. And he has a very touching argument that is theological in nature: animals cannot suffer, because pain is the result of original sin, and nowhere is it said that animals ate the forbidden fruit, and as a result, animals cannot suffer, it would be an injustice towards them because they did not commit this sin.[12] Only the human species can suffer. This is why we slice dogs in half and put them against

---

[11] Malebranche, The *Search after Truth*, Book VI, II part, chapter VII, Pleiades p. 467.

[12] See for example Malebranche's *The Search after Truth,* Book IV, chapter XI, Pleiades p. 717.

the barn door in order to watch the blood spill, this leads to the gentlemen of Port-Royal approving of vivisection, because animals don't suffer.

Bossuet

Amongst the authors who positioned themselves the most against Cartesianism, one finds Bossuet who fought to reconcile Descartes with Saint Thomas. There is no need to reduce Bossuet's meditation on this point. Bossuet went far enough and is proof of a great perspicacity and balance in this study. He said this: we are animals. Man is an animal. We have the experience of what is animal inside us and what comes from reflection and reason. The grandest, most complete being is man. And man is an animal. We can to a certain extent experience what it is to be animal. In a certain number of cases we are empirical and in those cases, we are animal. It is not impossible to experience, by way of inner meaning, what it is to be animal. This is more or less Bossuet's idea.

And what's more he says that the true problem is not asking if animals subsequently have an aim, congruence and reason in their behavior, because Bossuet says, the fact of having an aim, congruence, and reason is to a certain extent analogous to the order in the alignment of the organs of a living being. Specifically, he uses a quite tasty example, he says:

there is no order in the alignment of pomegranate seeds.[13] You know how in a pomegranate, the seeds are aligned in such a way; they are intertwined one to another to the extent that in certain areas of the pomegranate, there is no interstice between the two seeds. They are not rigorously regular in form, but they are so well fit into one another that there is absolutely no empty space from which one can separate them from each other and get them out of the pomegranate without smashing it. There is an order to the alignment of the seeds of the pomegranate, an organization of the anatomical type.

This anatomical organization in a plant is the same kind of organization that we call instinctive in the behavior of an animal that does one thing before doing another. This is the notion of structure. It is the notion of anatomical structure extended to the notion of the structure of behavior. The true problem is thus not knowing whether there is a structure, an aim, reason, links in animal behaviors, but knowing if the reason manifested by this aim, this organization is individually within them, or if it is found within the organization that made them. The question posed here is that of Creation itself. Does the animal species contain within in it that which pushes each individual to act in a certain way because

13 Bossuet, *Treatise on Free-Will*, ch. V, "The difference between man and beast".

it is a dog or a cat or a squirrel, in the same way pomegranate seeds are intertwined due to being in a pomegranate and grew in such a manner because it is in its anatomical nature to do so, or is it such that in each animal there is something that actually and actively constitutes the organizing principle, of the aim, reason, and links between the different actions? In other words, are we dealing with a specific activity or an individual activity? What is the carrier of reason? If it is the Creator who put reason in animals then it is a reason completely identical to that found in the pomegranate seeds, which are obviously specific. If it is an individual activity then it is similar to what is produced in a human being, of which it becomes the depositary to the extent a human being is a person, an individual, an organization of its actions and the correlations of its behaviors. This is how Bossuet poses the question without totally answering it. But he shows a clear awareness of what we could call the structure of behavior in correlation with the structure of organization at the anatomo-physiological level in living beings. Furthermore, already in Aristotle we find something partially of this type.

La Fontaine

But the one author who, in 16th century thought, took to the defense of the animal kingdom considering it to have been violated by systematic thinking and who

did it with a undeniable philosop
elements of positive science, e
as being the departure points fc
the study of customs, and anii
Fontaine. He is the first and de
thinker to engage in this type o
it is beginning in the 17ᵗʰ centur̄y ... uıcory of
animal behavior slowly but surely pried itself away
from philosophical theory and became a science of
experience, a matter of experience. This can be seen
most clearly in La Fontaine's fables such as "Address
to Madame de La Sablière". Here is an excerpt: "Now
you know, Iris, from certain sciences, that when an
animal thinks, the animal neither reflects upon an
object nor his thought". What he is getting at here is
that we concede that animals do not have reflexive
consciousness, reflection, in a certain manner
what we find in the *cogito*, which is the grasping
of the activity by itself. But this does not rule out
intelligence, reasoning, calculation and prediction.
Let's have a look at this "Address to Madame de La
Sablière", which is an important piece (there is also
the "Epistle to Madame de Montespan" which could
be relative to this point). It may be a somewhat boring
piece, but it strives at doing away with Cartesianism,
because Cartesianism is inadequate when it comes to
all vital phenomena.

w, generally speaking, La Fontaine's manner of
dering the genre of fables comes from Antiquity,
nd to this extent should not be considered by us as
a direct expression of the way La Fontaine studied
reality. The fable is a literary genre, but in the epistles
and discourses, he expresses his doctrine in a much
better manner, which is, as it were, a dissertation.

It is at the end of book IX. After the compliments
of the prevailing fashion, here is how things are
presented. He says (line 24 to 178):[14]

And thus it is, and take it not amiss
I mix with trifling fables, such as this,
A subtle bold philosophy.
(He's referring to Descartes' philosophy)

Which men call something new, and I
Know not if you have heard it, but they say
A beast is a machine which acts by springs,
With no more soul or will than lifeless things,
Like watches going blindly on their way.
(There is no prediction in the ticking hand
of the watch)

Open it, and look within
Wheels take the place of wit
One moves a second, that alike

---

14 The translation of these excerpts from La Fontaine have been updated from
the translation found on the website of the Association pour le musée Jean de
La Fontaine. http://www.la-fontaine-ch-thierry.net/assoc.htm

A third, and then we hear them strike.
And now the beast, as sages say,
Is moved precisely in this way.
'Tis stricken here, a neighboring spot
Receives the shock, till to the lot
(Here we see the theory of nervous conduction)

Of sense it comes at last;
And then the impression's fast.
But how? Why, by necessity, they say.
Passionless, will-less, without yea or nay,
The brute feels sorrow, joy, love, pleasure, pain,
Or what the crowd calls such, for 'tis in vain
To think it feels, a watch made with a spring.
And what are we? Oh quite a different thing.
Descartes, a mortal man whom the pagans
Had made a god, who holds the middle place
'Twixt man and spirit, as a donkey can
Hold his place 'twixt an oyster and a man
Descartes says I alone can think
Of all God's children, and I know
I think; the rest so far below
Myself, possess of thought no link.
Some say they think and can't reflect, but I
In them this thought deny.
This, Iris, you believe like me quite sound,
Yet when across the woods the noise of horn
And voice pursues the stag who fast is borne
Through tracks which he would oft in vain confound,
When he so full of years, a stag of ten,
Puts up a younger stag fresh prey, why then
He seems to reason to preserve his life;
His turns, his tricks, his changes, and his strife,

Each a great chief and better fate befits,
Yet his last honor is to be torn to bits.

So when the partridge with a mother's care
Spies danger for her brood which cannot fly,
She draws the dog's attention,
Feigning a broken wing, away from her progeny;
Then when the sportsman thinks
he has reached his prey,
Rises in the air and smiles, and says,
"Good-day."

Far in the north, by waters bound,
There exists a world
Where they say the population
Lives like those of earlier times
In a profound ignorance:
I am speaking of humans; as for the animals,
They are in the middle of construction work
Torrent wide to the opposite shore,
communicating across the banks.
The edifice is resistant and endures.
After plank of wood and mortar.
Each beaver acting in common on the task
The old making the young continue with rest
The master beaver conducts, holding his baton high.
Plato's Republic would merely be
the apprentice for this amphibious family.
They know how to build their homes in winter
They make bridges
Knowing the fruits of their art;
And after this, it seems somewhat wrong
To say beavers have no sense at all
All their knowledge until then given to swimming.

But here's another proof; a glorious king,
Defender of the North, told me a thing,
Which now for your instruction I recall.
That king beloved of victory, whose name
To the Ottoman Empire like a wall,
The Poles call Sobieski-can there fall
A lie from the royal mouth? The thought is shameful.
"Two kinds of beast on my frontiers live,"
He said, who in a hereditary feud
Fight like our generals with enduring skill:
Nay with more sill than Chan's poor men can give.
These funny animals, who are like foxes,
Have spies and watches, forts and sentry boxes,
Skirmish with guard advanced, and are well-versed
In all matters of that art accursed,
Mother of Heroes, daughter of the Styx.
To chant their many military tricks,
Hell must restore us Homer,
And also the rival of Epicurus!
What would the latter say of these examples?
All this nature's work, the work of springs,
That memory is bodily; many things
Besides those that facts explain away.
Animals have no need for it.
Once the object returned goes to the warehouse
To find, by the same path
The once traced image,
That returns by the same road,
Without the help of thought,
Causing the same event.
(This is an example of habit-memory)

With men, of course it's quite different,
Will determines us,

Not the object nor instinct. I speak, I walk;
Man feels a certain agent in him;
Everything obeys in my machine
This principle of intelligence.
It is distinct from the body, it clearly conceivable
More conceivable than the body itself:
Of all our movement it is the supreme arbiter.
But how does the body see it?
There's the point: I see a tool
Obeying the hand, but who guides it?
(This is example of the problem of the
communication of substances)

Who guides Heavens and their fast-moving paths?
Perhaps an angel is attached to these great bodies.
Some spirit lives in us and moves us and our hopes and fears.
But how? God knows. I don't.
And if there is need to speak without a lie,
Descartes is as ignorant as I.
We're equals here, for here we know nothing.
But what I know, Iris, is that in these animals I have named,
This spirit does not act, man alone is its temple.
But the beast must have some claim over the plant
Nevertheless the plant breaths, but who provide this a name?

The fable, *The Two Rats, The Fox and The Egg,* comes after and indicates the possibility to a certain extent of foresight (prevision) in the reasoning of animals. There is also another fable where La Fontaine wanted to directly attack the Cartesian doctrine. It's the fable of the owl ("The Mice and the Barn Owl"), that manifested a *pronoia*, principle of prevision, and calculation of what prevision allowed it to grasp. We find an old owl in a hollowed out tree. Old owls are always wiser than young owls, and we find the old owl, at the back of the hollowed out tree with "Mice without feet completely round and fattened up". The owl captured them during summer when mice were out and he reasoned that he should amputate them, and La Fontaine outlines the owl's reasoning for us: where a populace has feet, the populace flees. Consequently, if we cut the legs off of the mice and keep them in the tree, they will be nice fresh meat to eat in winter. But this populace will waste away, because having no more feet; they can no longer feed themselves. Thus, one must also gather grain, wheat, and the owl made provisions of a certain quantity of grain and wheat that he gave to the mice in order to keep them plump and fattened up. So, there you have several examples from La Fontaine where he tries to show not only do animals have a consciousness (he admits they don't have reflection), but there is consciousness because there are instances of individual organization and experience. We could also add (he almost adds this, he speaks of social animals, and I wonder if at that moment he was thinking about this), that there are cultural aspects in animals, what we could call within certain animal societies, culture. In particular, we have found that in certain societies of lions there are ways of hunting that are not shared practices in other lion societies. For example, one can look at how thirty or forty different animals species share a common practice of chasing their prey into a circle that closes in on them. These are not merely instinctive forms, but cultural forms. For example, a lion is raised in a group where this form of hunting is practiced. He knows only how to hunt based on this practice. It wouldn't appear that he practices this method of hunting out of instinct. We still are not sure whether there are very specific cultural phenomena practiced within animal societies. Here we have something that is almost implicit from what we have discussed about beavers, with their social hierarchy (a beaver directing all the others "holding the baton high in the air"). I have no idea if this is true, but this is the idea we can get from looking at the relationships of superiority-ascendancy between animals and their collective behavior.

Univocal Publishing
123 North 3rd Street, #202
Minneapolis, MN 55401
www.univocalpublishing.com

ISBN 9781937561017

Jason Wagner, Drew S. Burk
(Editors)
This work was composed in Berkely and Rockwell.
All materials were printed and bound
in November 2011 at Univocal's atelier
in Minneapolis, USA.

The paper is Mohawk Pure White Linen.
The letterpress cover was printed
on Crane's Lettra Fluorescent.
Both are archival quality and acid-free.